Michigan's State Stone

The Petoskey Stone

Written & Illustrated by:

Carrie Will

~Carrie Will~
Enjoy Gods Creation

Paperback: ISBN# 978-0-9972115-8-0

First Paperback Edition March 2025

Written and Illustrated by Carrie Will
Photography by Carrie Will and Gail Will

Cover & Interior Layout by Athena Publishing Group

Printed by LuLu Books

Publisher:

Athena Publishing Group
4629 Roehrs Rd, Beaverton, MI 48612

Publisher's Website: www.AthenaPublishingGroup.com

Introduction

I hope this book will inspire the reader to discover one of the greatest treasures that Michigan has.

It covers from the dinosaur age up to now, where you can explore. **ENJOY!**

Table of Contents

This book is for the ones who enjoy nature, want to learn history, and are curious about our state.
Not every state has a state stone;

Michigan has a unique one.

In this book, you will learn how God's creation has turned a living organism into a precious state Stone and a nature collector's treasure over many years.

We will begin with the Mesozoic Era, the time of the dinosaurs, and learn what has happened over time.

Our State Stone is very unique in how it was created.

Why is Michigan a popular destination, and where can we find this treature?

**Let's start
...at the beginning.**

A very long time ago . . .

Dinosaurs and corals existed together.

The warm sea was a perfect environment for the corals to live in; they were of different sizes and shapes.

The picture on the next page, shows the different colors and shapes.

The reef is a very colorful plant.

This picture shows the different vibrant colors of the reef and how it existed within its colony.

The tentacles gathered the food and the blue of the reef is the mouth.

Corals were many different colors and shapes (mainly hexagons). The colored area is the mouth; the white and green fingers are the tentacles, they work similar to our fingers.

The coral's fingers swayed in the warm water to gather food.

God made them special this way, by his design.

According to Pat Clark from Oregon State University, "We know the only thing changing in the Northern Hempisphere (20,000 years ago) were orbital changes."

"Melting in the North could have been triggered from the ice sheets having reached a size so big that the glaciers became unstable and therefore, the ice began to move life around in the oceans from the glaciers floating down from the Northern Hemisphere" (Biello, 2012).

Fossilized Dragonfly

Over the different eras,

the water temperature changed from warm to much colder.

Many plants and animals could not adjust well to the temperature change.

Some were able to move to a different home, but plants, such as corals, could not.

What is a fossil made from?

According to the Australian Museum, scientists define a fossil as,

"Any remains or traces of past life that are preserved. Fossils include plants, animal tissues, shells, teeth or bones, even bacteria!"

So, where does the Petoskey Stone come into all this science?

The Petoskey Stone is a fossilized coral!

What is in the middle of a Petoskey Stone?

Also known as the "eye" of the stone, fossilized corals.

The center of the fossilized corals were once the mouth when the corals were alive.

Now, the mouth of the corals have been fossilized which has made the Stone become darker in the middle, the white lines around the eye were once the tentacles.

Hexagonaria

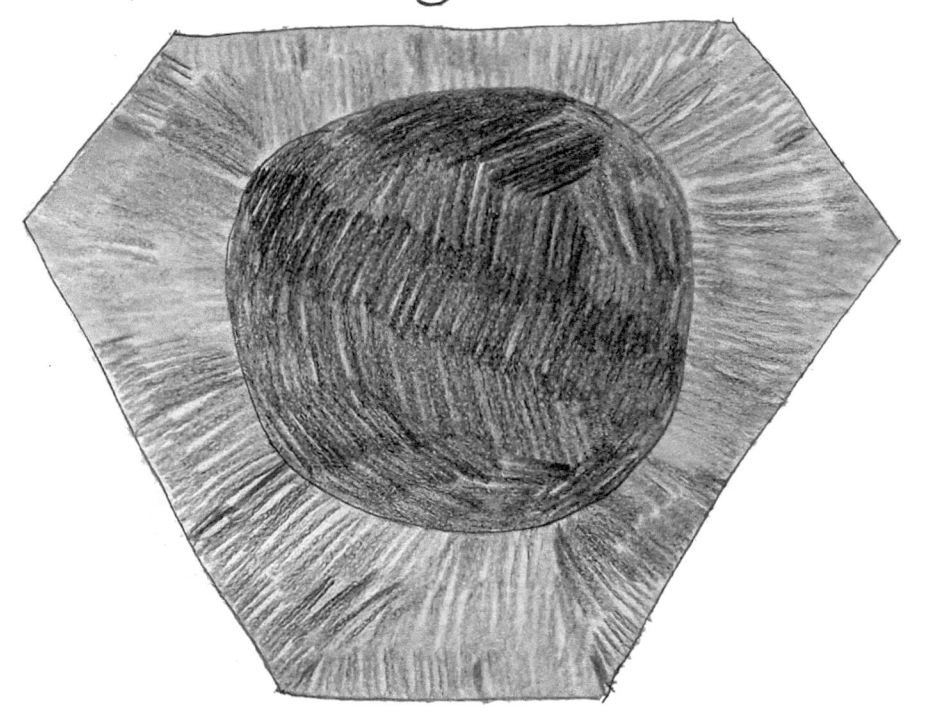

~~ Petoskey Stone ~~

Hexagonaria Percarinata,

is the scientific name of the Petoskey Stone.
Does the shape look familiar - Hexagon, 6-sided?

The shape is developed over the process of being
tumbled in the water and abrasive obstacles that
helped to form and polish the Petoskey Stone.

Devonian Era Fossil Coral

During the Devonian era,

Rugose Corals

lived and thrived

in the warm sea waters

of North America!

Along   A   Lakeshore

Petoskey Stones can be found along a lakeshore, in a field, a pile of rocks, but most commonly along the

Great Lakes shorelines of Northern Michigan.

Turn your learning into a Scavenger Hunt with your family and friends!

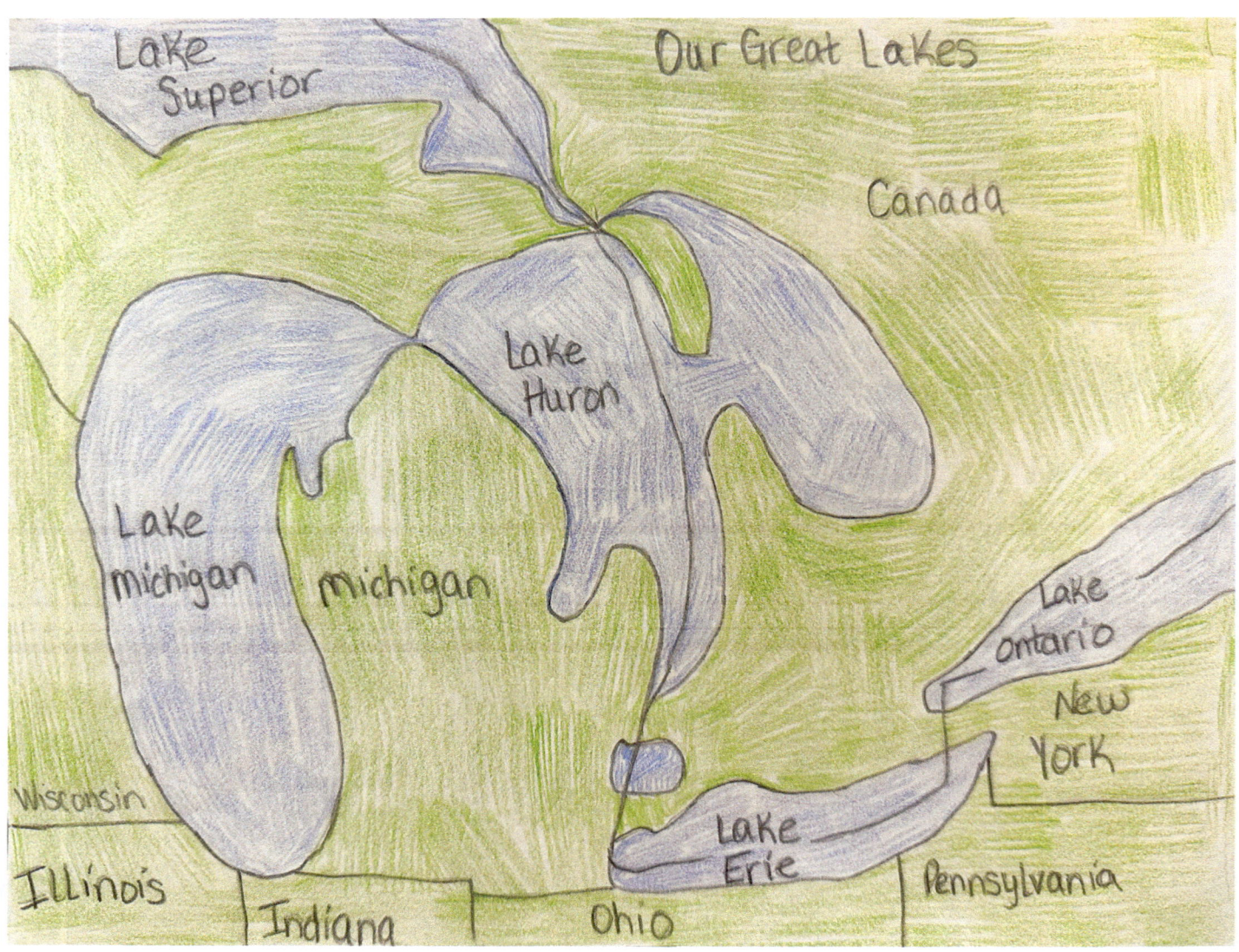

Our Great Lakes

Lake Superior

Canada

Lake Huron

Lake michigan

michigan

Lake ontario

New York

Wisconsin

Illinois

Indiana

Ohio

Lake Erie

Pennsylvania

Over many, many years the "Earth's tectonic plates began shifting enough to push Michigan up to the 45th parallel.

As this was happening it was pushing the fossils to be primarily across the lower peninsula of Michigan.

Today, this has become a highlight attraction to the Petoskey" area of Northern Michigan!

Where does Petoskey originate from?

Petoskey originates from a Great Ottawa Chief,

"Native American name Pet-O-Sega, which translates to "Rays of Dawn" or "Rising Sun."

Over the years, legends have changed and stories have been shared about how long the Petoskey name has been around.

Many historians and scientists believe it originated in the late 18th Century.

Legend takes us back to the Ottawa Indian Tribe.

From legend, a French man named Antoine Carre made a visit to the area that is now known as Petoskey.

Overtime, he met then married an Ottawa princess, they had a son and he then became

the Chief Petosegay.

June

Public
Act
89

Official
State
Stone

1965

In 1965, Governor George Romney signed the House Bill 2297, which made the Petoskey Stone Michigan's State Stone.

Therefore, making this Stone very famous to bringing tourists to Michigan to find this unique treasure!

Fisherman's Island, located in Cherlevoix Michigan, also is located on Lake Michigan (and Lake Huron), are very popular destinations for

Petoskey stone hunters, such as us!

While looking for stones and treasures, when you look up, look for Skydivers!

Petoskey Stone Hunting Tools

Wading Boots

Sifter

Bucket for Stones

Here are a few tools and supplies that will help you along in your adventures.

Wading Boots

will help keep your feet dry from the cool water of Lake Michigan.

Long Handled Sifter

to help sort the sand and stones.

Small Bucket

The bucket will get heavy too! (State law states to only take 25 pounds of Petoskey stones per year!)

My goal with this book was to teach you about
the Petoskey Stone from start to finish.

My hope is that you go out
        and find a few yourself,
            take pictures, and

enjoy our Great Lakes!

Resources

Austrialian Museum. 2024 www.austrialian.museum.com

Biello, David. April 4, 2024. What Thawed The Last Ice Age. SCIAM

Kchodl, Joseph J. "Paleo Joe". Mackinaw Center for Public Policy. 'Petoskey Stone': Michigan's State Stone. August 17, 2011. www.mackinaw.com

The Petoskey Stone. www.michigan.gov

The Hidden Legend Behind the Petoskey Stone That Everyone Should Know. Michigan Nail Polish. October 2, 2020. https://www.northernnailpolish.com

About the Author & Illustrator!

Carrie Will

Hi, I am Carrie Will, the author of Michigan's State Stone. I live in Gladwin, Michigan and am the Executive Director for the Gladwin County Chamber of Commerce. I have an Associate's degree in Biology, Bachelor of Science degree in Business Management and a Master's degree in Human Resources.

Three years ago, I was working for a corporation in Charlevoix Michigan, and I would go to Fisherman's Island when I had a little extra time. I did not know at the time what a Petoskey stone was, I had never found one for myself.

One day as I was walking along the Lake Michigan shoreline, a little boy came up to me and asked me if he had a Petoskey stone, I said yes and good job for finding one. His mother then asked me if I knew how the Petoskey stone came to be, I told her I did not, but that would be interesting to learn. So here we are today with this book.

I hope that this book will encourage us all to get outside, find your own Petoskey stone and enjoy what God has given us!

"Thank you mom, Gail Will,

for taking my picture."